# Workbook 1

William Collins' dream of knowledge for all began with the publication of his first book in 1819. A self-educated mill worker, he not only enriched millions of lives, but also founded a flourishing publishing house. Today, staying true to this spirit, Collins books are packed with inspiration, innovation and practical expertise. They place you at the centre of a world of possibility and give you exactly what you need to explore it.

Collins. Freedom to teach.

Published by Collins
An imprint of HarperCollins*Publishers* Ltd.
The News Building
1 London Bridge Street
London
SE1 9GF

HarperCollins*Publishers*
1st Floor
Watermarque Building
Ringsend Road
Dublin 4
Ireland

**Browse the complete Collins catalogue at www.collins.co.uk**



10 9 8 7 6 5

ISBN: 978-0-00-836893-7

Second edition

Contributing authors: Phillipa Skillicorn, Karen Morrison, Tracey Baxter, Sunetra Berry, Pat Dower, Helen Harden, Pauline Hannigan, Anita Loughrey, Emily Miller, Jonathan Miller, Anne Pilling, Pete Robinson.



British Library Cataloguing in Publication Data
A Catalogue record for this publication is available from the British Library.

Commissioning editor: Joanna Ramsay
Product manager: Letitia Luff
Development editor: Karen Williams
Project manager: 2Hoots Publishing Services Ltd
Proofreader: Caroline Low
Cover designer: Gordon MacGilp
Cover illustrator: Ann Paganuzzi
Image researcher: Emily Hooton
Illustrators: Beehive Illustration (John Batten, Moreno Chiacchiera, Phil Garner, Kevin Hopgood, Tamara Joubert, Simon Rumble, Jorge Santillan, Matt Ward); Graham-Cameron Illustration (Sue Woollatt)
Design and typesetting: Ken Vail Graphic Design Ltd
Production controller: Lyndsey Rogers
Printed and bound by: Grafica Veneta S.p.A, Trebaseleghe (PD)

With thanks to the following teachers and schools for reviewing materials in development: Preeti Roychoudhury, Sharmila Majumdar and Sujata Ahuja, Calcutta International School; Hawar International School; Melissa Brobst, International School Budapest; Rafaella Alexandrou, Diana Dajani, Sophia Ashiotou and Adrienne Enotiadou, Pascal Primary School Lefkosia; Niki Tzorzis, Pascal Primary School Lemesos; Vijayalakshmi Chillarige, Manthan International School; Taman Rama Intercultural School.

**Acknowledgements**
The publishers wish to thank the following for permission to reproduce photographs. Every effort has been made to trace copyright holders and to obtain their permission for the use of copyright materials. The publishers will gladly receive any information enabling them to rectify any error or omission at the first opportunity.

p21tl Filip Fuxa/Shutterstock, p21tr mj007/Shutterstock, p21cl Angelo Ferraris/Shutterstock, p21cr Kostiantyn Karpenko/Shutterstock, p21bl Mariyana Misaleva/Shutterstock, p21br Paul Cowan/Shutterstock, p27a sagir/Shutterstock, p27b Elnur/Shutterstock, p27c Dragan Milovanovic/Shutterstock, p27d Skylines/Shutterstock, p27e Picsfive/Shutterstock, p27f Robbi/Shutterstock, p73tl Somchai Som/Shutterstock, p77tcl Kitch Bain/Shutterstock, p77tcr zentilia/Shutterstock, p77tr Mile Atanasov/Shutterstock, p77bl lucadp/Shutterstock, p77br viphotos/Shutterstock.

# Contents

## Topic 1 Plants



## Topic 2 Humans and other animals



## Topic 3 Materials

## Topic 4 Forces and sound



## Topic 5 Electricity and magnetism



## Topic 6 Earth and Space

Student's Book p 2



# Making predictions

Look at the pictures. What do you think will happen next? Why do you think this?

Write your predictions. Remember to describe your predictions well.

| | |
|---|---|
|  | *I think* ______________________<br>______________________<br>*because* ______________________<br>______________________ |
|  | ______________________<br>______________________<br>______________________<br>______________________ |
| 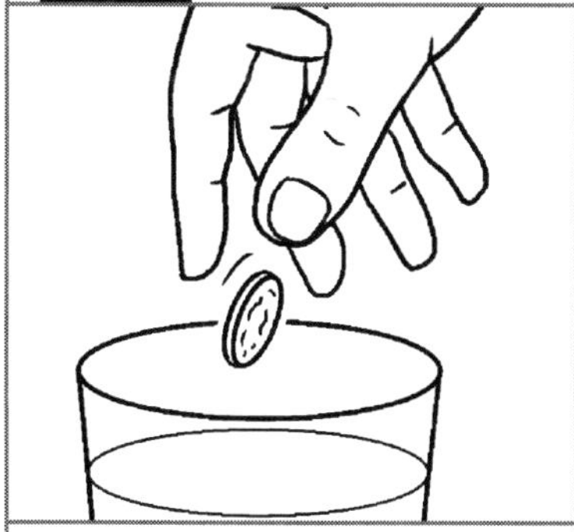 | ______________________<br>______________________<br>______________________<br>______________________ |
|  | ______________________<br>______________________<br>______________________<br>______________________ |

# Do plants need water?

I think the plant that had water will look like this:

I think the plant that had no water will look like this:

Circle the correct answers.

**1** I think the plant that was watered will grow. Yes No

**2** I think the plant that was not watered will grow. Yes No

Look at the different plants. Were your predictions correct?

Yes ☐ No ☐ Not sure ☐

Do plants need water?

Student's Book p 4

1.2 Is it alive?


# Is it living or non-living?

Look at the pictures of living and non-living things.

Circle all the living things.

Student's Book p 6

1.3 Plants and animals are living things

# Is it a plant, an animal or non-living?

Look at the pictures. Circle the plants. Underline the animals.
Put a cross through the non-living things.

Student's Book p 8

1.4 Things that have never been alive

# Has it ever been alive?

Look at the pictures. Circle the things that have never been alive.

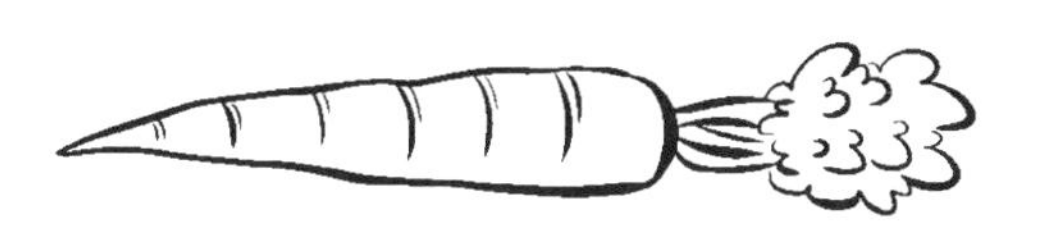

Student's Book p **8**

**1.4** Things that have never been alive

# My pictures

Draw one living thing and one thing that has never been alive.

This is **living**.

This has **never been alive**.

Student's Book p 10
1.5 Parts of a plant

# Differences in plants

Choose two plants near your school or home.

Say where the plants grow.

Draw or write about the differences between the plants.

| Plant 1 | Plant 2 |
|---|---|
| Leaves | Leaves |
| Stem | Stem |
| Flower | Flower |

Student's Book p 10
1.5 Parts of a plant

# Plant parts

Draw a flowering plant, showing the main parts.

Label your drawing with these words.

stem  root  flower  leaf

Student's Book p 12
1.6 What do plants need to survive?

# Does light affect plant growth? (1)

Draw a picture of the healthy plant here.

I predict that the plant that did not get light will look like this:

**Student's Book p 12**
**1.6** What do plants need to survive?

# Does light affect plant growth? (2)

Draw a picture of the plant that did not get light here.

Look back at your prediction. Were you correct?

Yes ☐ No ☐ Not sure ☐

*The plant grew like this because* ______________________________

______________________________________________

______________________________________________

Student's Book p 12
1.6 What do plants need to survive?

# What do plants need?

Use words from the box to complete the passage.

| water | leaves | light | flowers | roots |
|---|---|---|---|---|

All plants have some parts that are the same.

Plants have ______________ and ______________.

Plants need ______________ to live and grow well.

Plants need ______________ to live and grow well.

Some plants also have ______________.

**Student's Book p 14**
Looking back **Topic 1**

# My favourite plant

Draw a picture of your favourite plant. Label your drawing with these words.

stem root flower leaf

My favourite plant needs ______________ and ______________ to live and grow well.

Student's Book p 16
2.1 Parts of the human body

# My body

Draw a picture of yourself. Label the different parts of your body.

head neck arm hand leg stomach
foot chest shoulder finger toe

Student's Book p 18
2.2 Our senses

# Which sense organ?

Draw a line to match each sense to the correct sense organ.

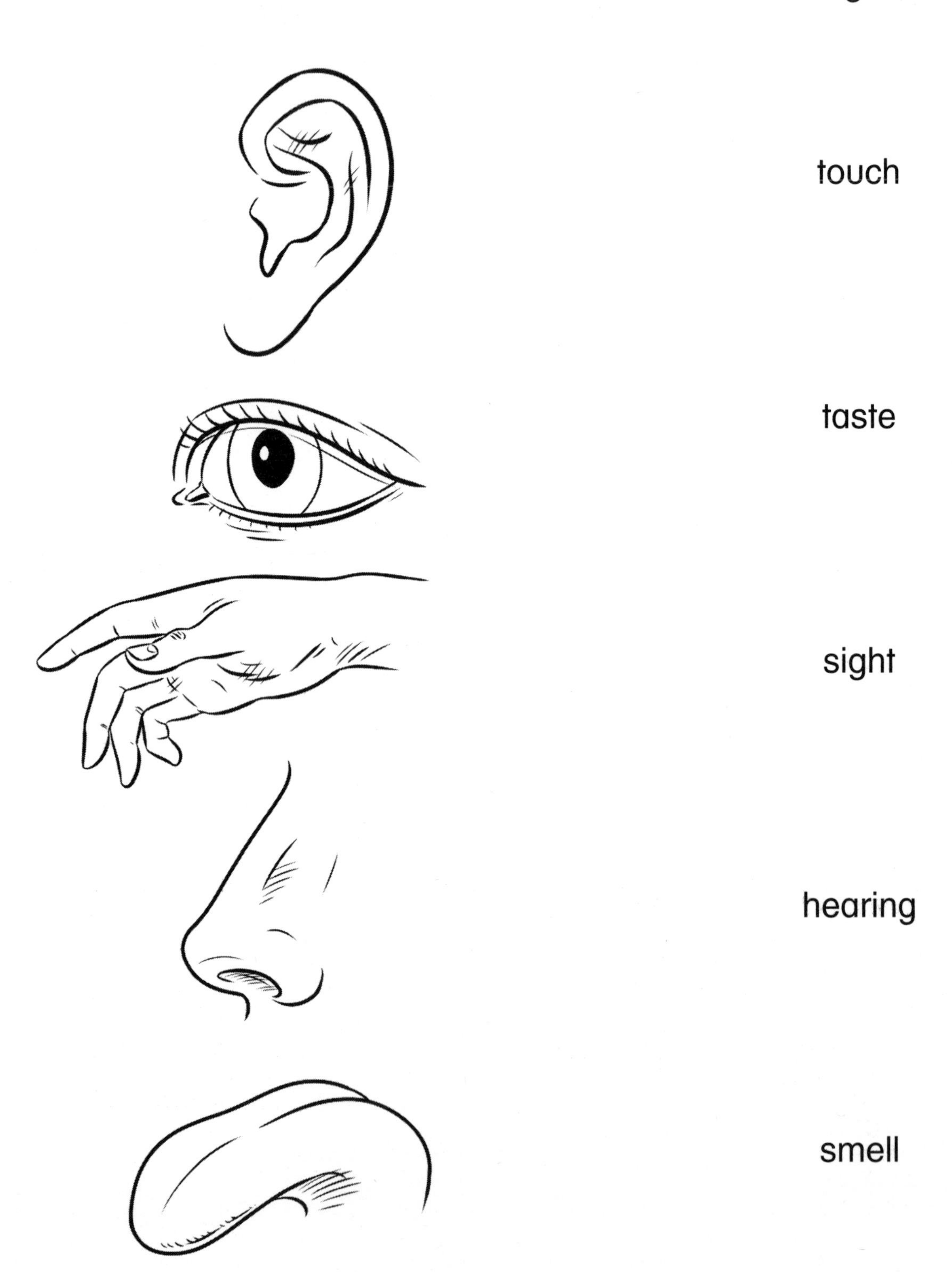

Student's Book p 18
2.2 Our senses

# My senses

Draw a line from each sentence to the correct sense organ.

Then use the words below to complete the sentences.

| sight | hearing | smell | touch | taste |
|---|---|---|---|---|

I use my sense of ________ to find out what something smells like.

I use my sense of ________ to feel different textures.

I use my sense of ________ to know if something is sweet or not.

Student's Book p 20
2.3 Using our senses

# Listening

Listen to some different sounds. Draw or write what you hear.
Use this table.

| Sound 1 | Sound 2 | Sound 3 | Sound 4 |
|---|---|---|---|
| | | | |

| Sound 5 | Sound 6 | Sound 7 | Sound 8 |
|---|---|---|---|
| | | | |

Student's Book p 20
2.3 Using our senses

# Which sense would you use?

Tick (✔) the sense that you would use in each situation.

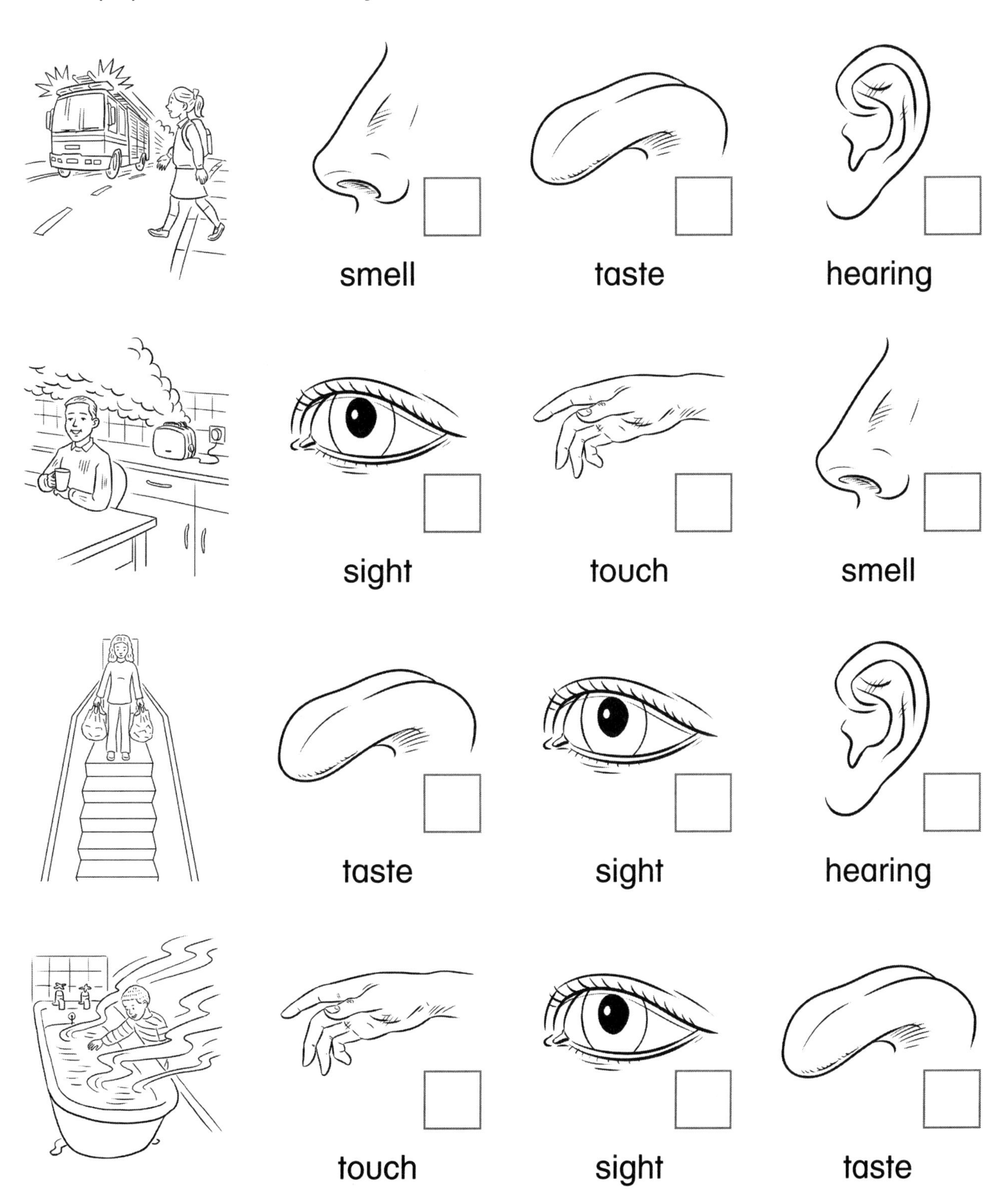

Student's Book p 22
2.4 Users of science

# What do we see?

Look at the pictures.

Circle all the things that we enjoy with our sense of sight.

Student's Book p 22



# Investigating sight (1)

Draw two things that you will try to do with a blindfold.

*I think I will find* ______________________________

______________________________

______________________________

*I think this because* ______________________________

______________________________

______________________________

Student's Book p 22
2.4 Users of science

# Investigating sight (2)

Draw or write what happened here.

Did wearing the blindfold make a difference?

Yes ☐ No ☐

*I found* ______________________________

______________________________

______________________________

______________________________

Look back at your predictions. Were you correct?

Yes ☐ No ☐ Not sure ☐

Student's Book p 24
2.5 What do animals need to survive?

# Is it safe to drink?

Tick (✓) the box if you think the water is safe to drink.

| | |
|---|---|
|  Yes ☐ No ☐ |  Yes ☐ No ☐ |
|  Yes ☐ No ☐ |  Yes ☐ No ☐ |
|  Yes ☐ No ☐ |  Yes ☐ No ☐ |

Student's Book p 28
2.7 Humans are different

# Hand spans

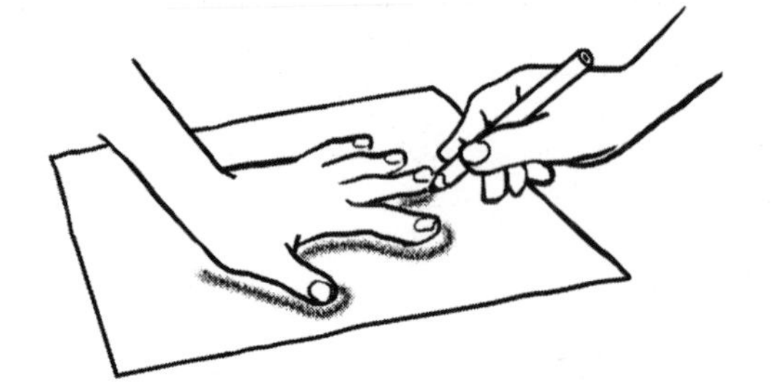

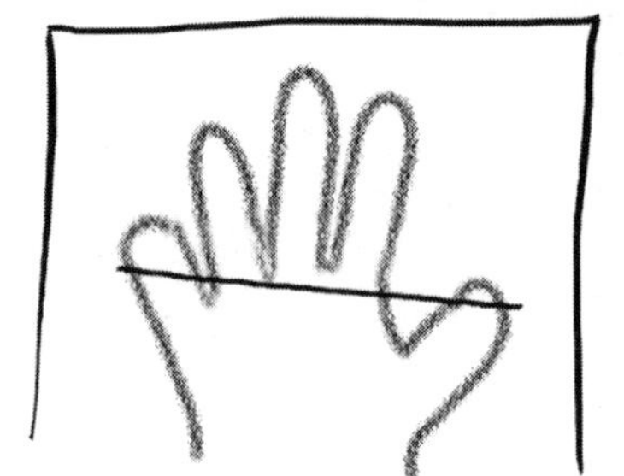

Put your hand in the box. Draw round your hand.

Draw a line across your hand span. Measure the line. ____________

Compare your hand span with the rest of your group.

Who has the biggest hand span? ____________________________

Who has the smallest hand span? ____________________________

Does the oldest person have the biggest hand span? ____________

Student's Book p 30
Looking back Topic 2

# Senses at the fun fair

The children can see ______________________________

______________________________

______________________________

The children can hear ______________________________

______________________________

______________________________

The children can smell ______________________________

______________________________

______________________________

Student's Book p 32
3.1 Similar or different?

# Materials sort

Sort the objects into the right circles.

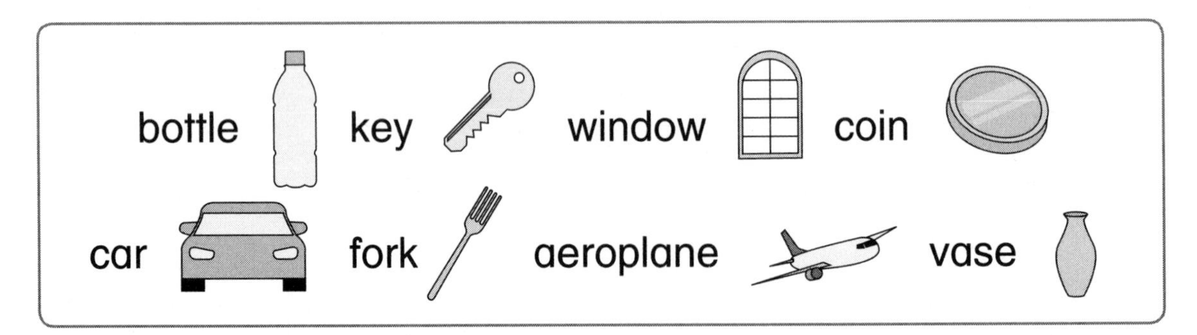


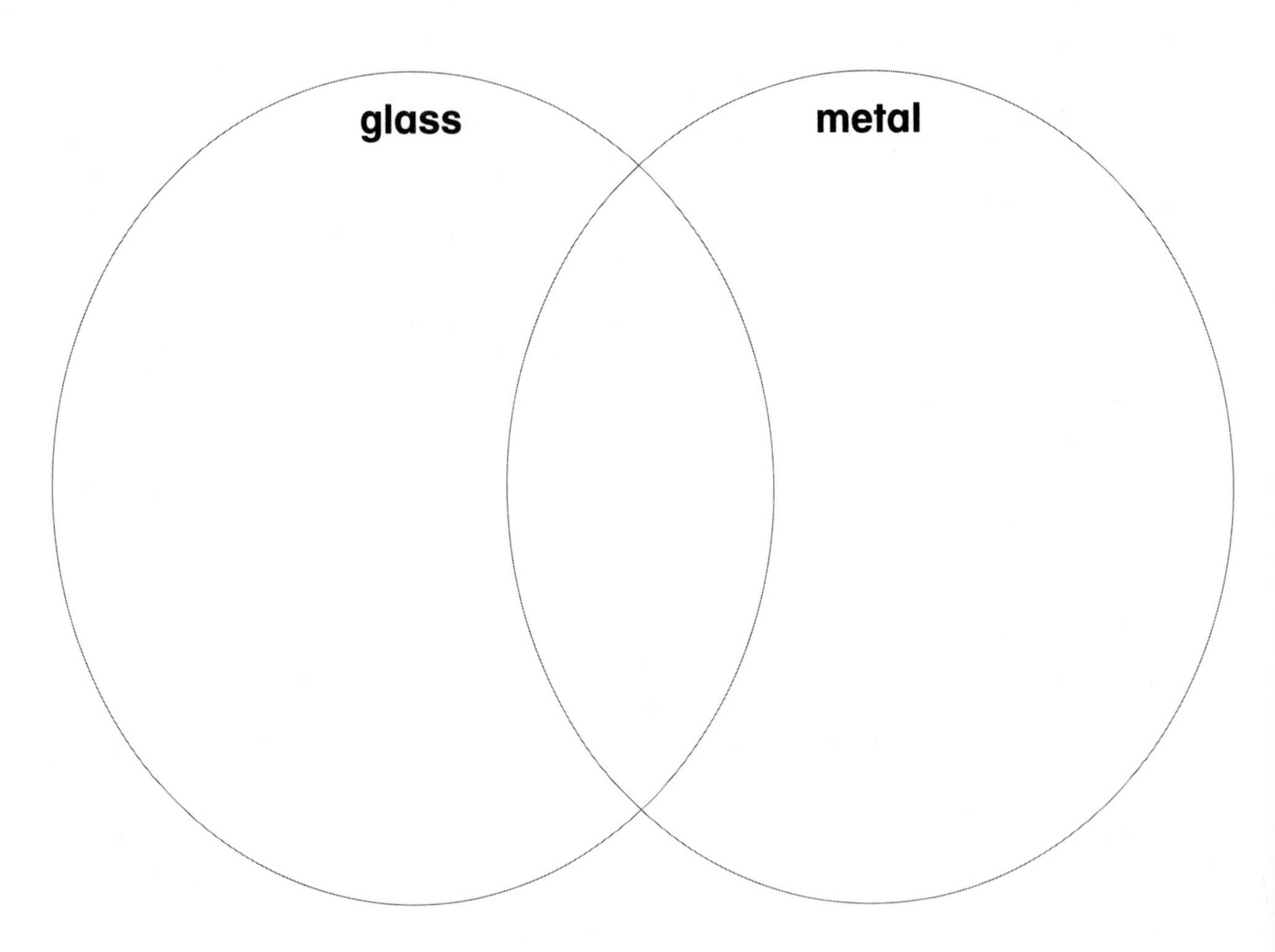


Which objects can go in both groups?

---

Student's Book p 34



# Comparing materials

Investigate the properties of two objects.

Record your findings in the table.

| Property | Plastic spoon | Wooden spoon |
|---|---|---|
| Is the object hard? | yes | |
| Is the object heavy? | | |
| Is the object smooth? | | |
| Can you bend the object? | | |
| Can you break the object? | | |
| Is the object shiny? | | |
| Can you see through the object? | | |
| What colour is the object? | | |

Student's Book p 34

3.2 Properties of materials

# Strongest or weakest?

1 Before you begin, predict which material you think will be the strongest and which will be the weakest.

________________________________

2 How many weights did you put in each bag before the material broke?

| Material | Number of weights |
|---|---|
| | |
| | |
| | |
| | |
| | |

3 Which material was the strongest? ________________

4 Which material was the weakest? ________________

5 Were your predictions correct? ________________

6 What did you do to make the investigation fair? ________________

________________________________

Student's Book p 36



# What are its properties?

Choose the properties that describe each object.

hard soft light heavy strong smooth
rough shiny waterproof flexible absorbent

| Object | Properties |
|---|---|
| cotton shirt | |
| brick | |
| rubber gloves | |
| newspaper | |
| metal foil | |
| plastic | |

Student's Book p **36**
**3.3** More properties

# Absorbent materials

Choose three absorbent materials.
Draw and label a picture of each material.

Can you name any more absorbent materials?

# Exploring different objects

Choose three different objects. Draw and describe each object. How does the object look? How does the object feel? What properties does each object have?
Talk about how you could find the answers to these questions.

| | |
|---|---|
| **Object 1** | |
| **Object 2** | |
| **Object 3** | |

Student's Book p 38
3.4 What material is it?

# What material is it made from?

Tick (✓) the material that each object is made from.

| | |
|---|---|
| **1** The chair is made from ...<br>wood ☐<br>paper ☐<br>cotton ☐ | **2** The window is made from ...<br>stone ☐<br>wool ☐<br>glass ☐ |
| **3** The bridge is made from ...<br>cardboard ☐<br>fabric ☐<br>metal ☐ | **4** The dress is made from ...<br>glass ☐<br>fabric ☐<br>metal ☐ |
| **5** The bag is made from ...<br>wood ☐<br>stone ☐<br>paper ☐ | **6** The hammer is made from ...<br>cotton ☐<br>metal ☐<br>glass ☐ |

Student's Book p 38
3.4 What material is it?

# Materials hunt

Find some objects made from wood, metal, fabric and plastic.

| Object | What is the object made from? | Tick when you have found the object or a similar item. |
| --- | --- | --- |
| paperclip | | |
| scarf | | |
| bottle | | |
| pencil | | |
| shirt | | |
| spoon | | |

Student's Book p **40**
**3.5** More materials

# Materials at home

Draw three things made from different materials at home.

Write the names of the materials under each picture.

Describe the properties of each material.

# Sorting materials (1)

Cut out pictures of objects made from wood and metal.

Glue your pictures here.

**Wood**

**Metal**

Student's Book p 42
3.6 Sorting materials

# Sorting materials (2)

Cut out pictures of objects made from plastic and fabric.

Glue your pictures here.

**Plastic**

**Fabric**

# Odd one out

Which is the odd one out? Explain why.

paper stone cardboard

*The odd one out is ____________ because ____________*

*______________________________.*

saucepan candle key

*The odd one out is ____________ because ____________*

*______________________________.*

wool cotton wood

*The odd one out is ____________ because ____________*

*______________________________.*

car book magazine

*The odd one out is ____________ because ____________*

*______________________________.*

Student's Book p 44
3.7 Making smaller groups

# Investigating paper (1)

Look at and feel six different types of paper.
Write the name and properties of each type.

| Paper | Name | Properties |
|---|---|---|
| 1 | | |
| 2 | | |
| 3 | | |
| 4 | | |
| 5 | | |
| 6 | | |

Paper _______ is the strongest.

Paper _______ is the thinnest.

Paper _______ is the smoothest.

Paper _______ is the heaviest.

Paper _______ is the shiniest.

Paper _______ is the most flexible.

Student's Book p 44

3.7 Making smaller groups

# Investigating paper (2)

Glue the six different types of paper here.

| Paper 1 | Paper 2 |
| --- | --- |
| Paper 3 | Paper 4 |
| Paper 5 | Paper 6 |

Student's Book p 44
3.7 Making smaller groups

# Investigating plastic

Draw and label the plastic that was the easiest to see through.

Draw and label the plastic that was the easiest to stretch.

Draw and label the plastic that was the easiest to break.

Student's Book p 46

3.8 Materials can change shape

# Modelling clay shapes

**1** Make four pencil shapes from clay.

- Squash one.
- Twist one.
- Bend one.
- Stretch one.

**2** Draw the shapes you made.

| Clay after I squashed it: | Clay after I bent it: |
|---|---|
| Clay after I twisted it: | Clay after I stretched it: |

**3** Can you change the shape of a wooden block by squashing, bending, twisting or stretching the block? ________

**4** Complete these sentences.

*I can change the shape of the clay because* ________________

________________________________

*I cannot change the shape of the wood because* ________________

________________________________

Student's Book p 46
3.8 Materials can change shape

# Changing the shape of materials

**1** What do you want to find out?

______________________________________________

______________________________________________

**2** What will you do to find this out?

______________________________________________

______________________________________________

______________________________________________

**3** Complete this table to show your findings. Tick (✓) the columns.

| Material tested | Can be squashed | Can be bent | Can be twisted | Can be stretched |
|---|---|---|---|---|
| | | | | |
| | | | | |
| | | | | |
| | | | | |
| | | | | |

**4** What did you learn about materials from this test?

______________________________________________

______________________________________________

Student's Book p 48
3.9 Squashing and bending

# Can you squash it?

Can you squash each object?

Predict what you think will happen.

Test each object to see if your predictions were correct.

| Object | Will the object squash? | What happened? |
| --- | --- | --- |
| | | |
| | | |
| | | |
| | | |
| | | |

Student's Book p 48
3.9 Squashing and bending

# How easy is it to squash?

You are going to investigate how easy it is to squash two different types of balls.

Draw what you think will happen to the shape of each ball when you try to squash the ball.

| Tennis ball | Cricket ball |
| --- | --- |
| | |

Why do you think this?

______________________________

______________________________

Do a test to find out if your predictions were correct.

Draw or write what happened here.

| |
| --- |
| |

Were your predictions correct?

Yes ☐ No ☐ Not sure ☐

Student's Book p 50
3.10 Stretching and twisting

# Testing for stretching

Trang tested five materials to see which material stretched the most. Each piece of material was the same size at the start. These are her results:

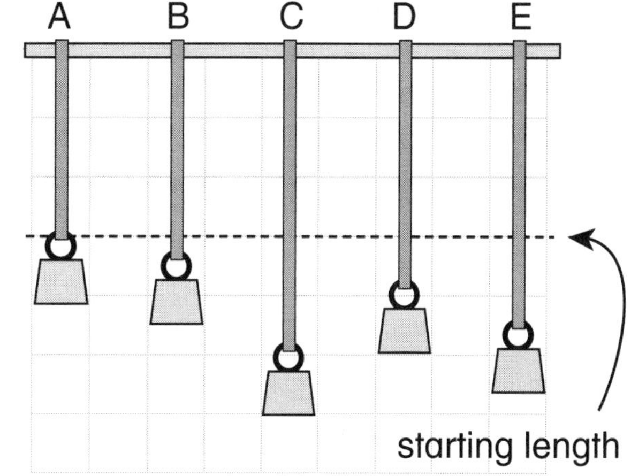


**1** Which material stretched the most?

---

**2** Which material did not stretch?

---

**3** Rank the materials in order from the most stretchy to the least stretchy. Write A–E in the table.

| | **Material** |
|---|---|
| Most stretchy | |
| ↓ | |
| | |
| | |
| Least stretchy | |

**4** One of the materials was wood.

Which material do you think this was? Why do you think this?

---

---

---

Student's Book p **50**
**3.10** Stretching and twisting

# Stretching and twisting

Stretch your clay into a pencil shape.

Gently pull on both ends.

Draw your shape.

What is the longest shape you can make without breaking the clay?

Twist your pencil shape.

Draw your shape.

How many times can you twist your shape before it breaks?

# Twisting test

Make a thin pencil shape and a thick pencil shape.

Predict which shape you can twist more times before the clay breaks. Tick (✓) the box.

☐ the thin pencil shape ☐ the thick pencil shape

Draw or write what happened here.

Why do you think this happened?

_______________________________________________

_______________________________________________

Was your prediction correct?

Yes ☐ No ☐ Not sure ☐

Student's Book p 52
3.11 Uses of science

# Wood and stone

Look at some pictures of things made from wood and stone.

Choose one thing made from each material.

Stick the pictures in the box and write the name of the object.

Describe the features of each material.

| | |
|---|---|
| Object: ______________ | The material is ______________<br>______________<br>______________<br>______________<br>______________<br>______________ |
| Object: ______________ | The material is ______________<br>______________<br>______________<br>______________<br>______________<br>______________ |

# Materials in the classroom

Look around your classroom.

Draw and label three things made from different materials.

Write the name of the material under each object.

What does each thing do?

1 ______

2 ______

3 ______

Describe the properties of the materials.

1 ______

2 ______

3 ______

Student's Book p **54**
Looking back **Topic 3**

# Sports equipment

Think about the different sports equipment.
What material is each object made from?
What properties does each object have?

| Object | Material | Properties |
|---|---|---|
| tennis racket strings | plastic | quite flexible and strong |
| cricket bat | | |
| tennis ball | | |
| weights | | |
| trainers | | |

Student's Book p 56

4.1 Thinking and working scientifically

# Investigating cars (1)

You are going to investigate which toy car moves the furthest.

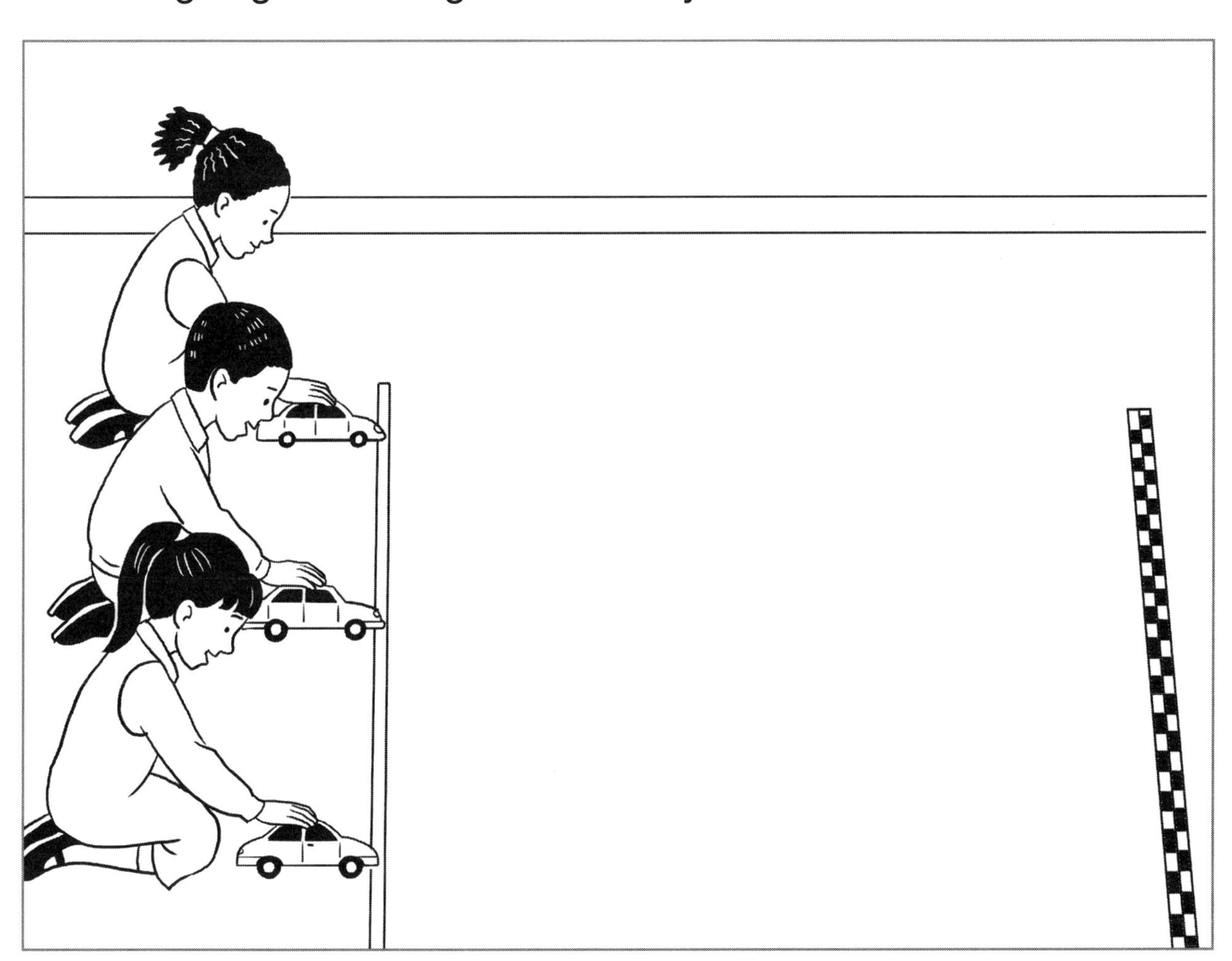

Write your predictions here.

*The car that will move the furthest is* ______________________.

*I think this because* ______________________

______________________

______________________.

Student's Book p 56
4.1 Thinking and working scientifically

# Investigating cars (2)

Draw the position of each car after it was pushed.

Measure how far each car travelled.

| | Distance travelled |
|---|---|
| Car 1 – small push | |
| Car 2 – medium push | |
| Car 3 – big push | |

Which car won the race? ______________________

Which car was in second place? ______________________

Which car was in third place? ______________________

Which car moved the furthest? ______________________
Explain your results.

*The winning car moved the furthest because* ______________________

______________________________________________

______________________________________________.

Look back at your prediction. Were you correct?

Yes ☐ No ☐ Not sure ☐

Student's Book p 58
4.2 Movement

# Movement survey

Make a list of the different animals you see near your school. Tick (✓) to show how they move.

| Name of animal | Walks | Flies | Crawls | Other |
|---|---|---|---|---|
| cat | ✓ | | | |
| | | | | |
| | | | | |
| | | | | |
| | | | | |
| | | | | |
| | | | | |
| | | | | |

I saw _______________ animals that walk.

I saw _______________ animals that fly.

I saw _______________ animals that crawl.

The most common type of movement was _______________.

Student's Book p 58
4.2 Movement

# Different ways of moving

Draw pictures to show four things that move in different ways. The things can be living or non-living.

| Roll | Slide |
| --- | --- |
| | |

| Fly | Bounce |
| --- | --- |
| | |

Student's Book p 60
4.3 Pushing and pulling

# Which ones show a push?

Colour the pictures that show a push.

Student's Book p 60
4.3 Pushing and pulling

# Toys

Draw a toy in each circle. Draw one push toy and one pull toy.

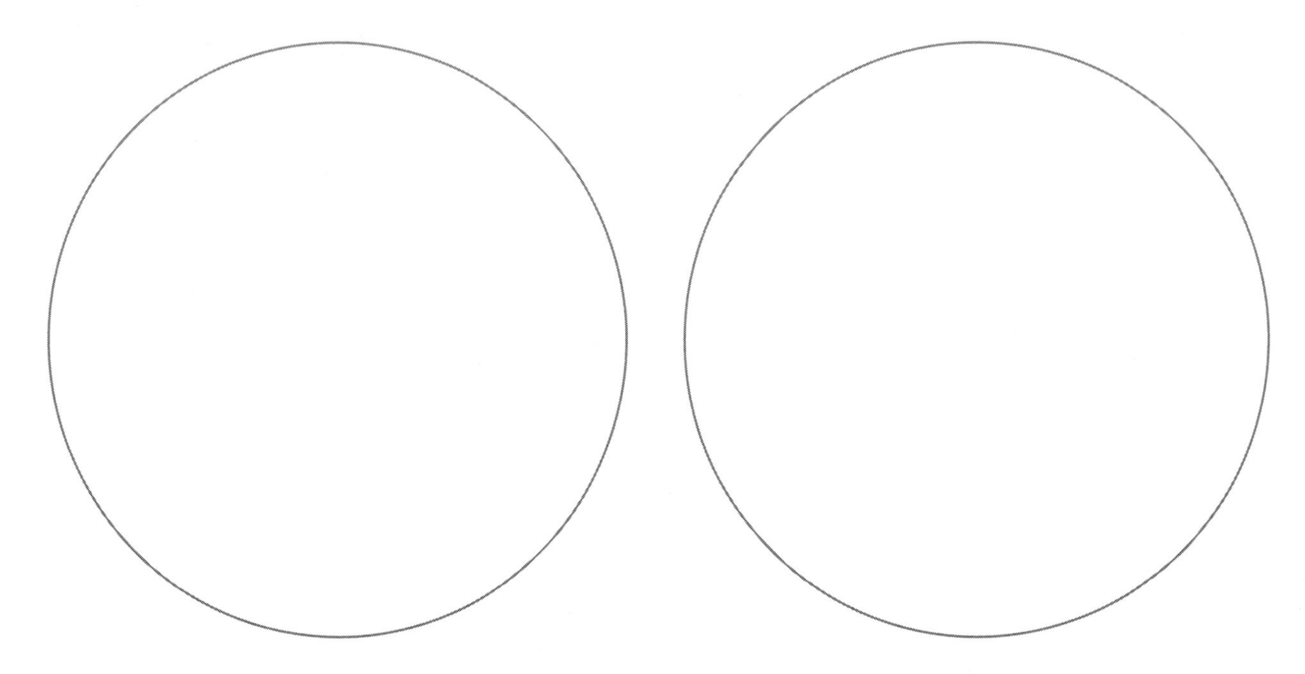

To make the ________________ work you need to ________ it.

This is what happens ____________________________________________

________________________________________________________________

________________________________________________________________.

To make the ________________ work you need to ________ it.

This is what happens ____________________________________________

________________________________________________________________

________________________________________________________________.

Student's Book p 62
4.4 Pushes and pulls

# Pushing or pulling?

Look at the picture of a bathroom.

Put a red circle around the things that work by pushing.

Put a blue circle around the things that work by pulling.

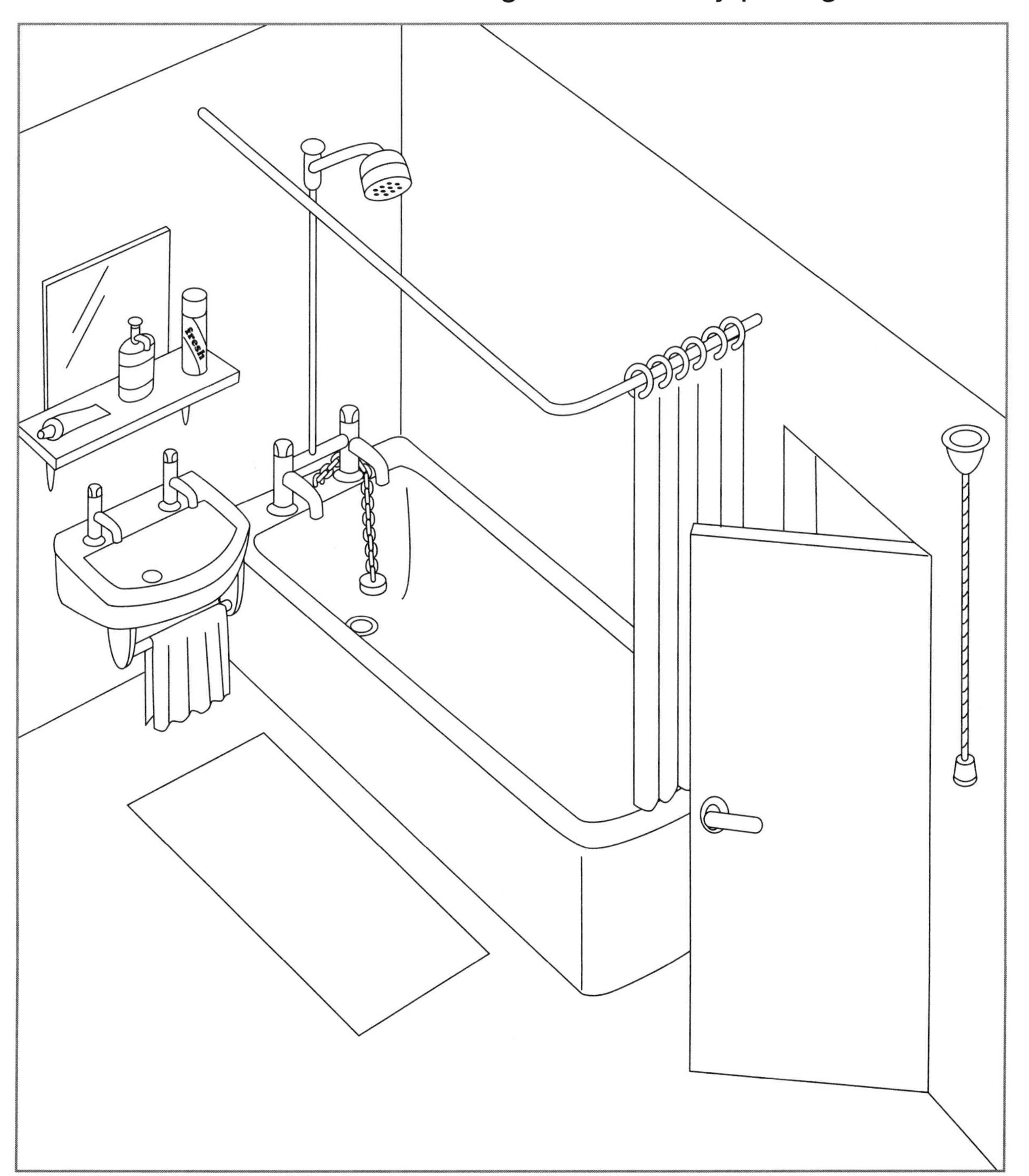

Student's Book p 62
4.4 Pushes and pulls

# Blowing balls

Draw a picture of the balls you think will be the easiest and the hardest to move by blowing.

| Easiest | Hardest |
| --- | --- |
| | |

Test the balls and then draw a picture to show your results.

Was your prediction correct? ____________________

Student's Book p 64
4.5 Floating and sinking

# Will it float?

Draw each object and describe its shape.

Predict if each object will float or sink.

Test each object and record your findings in the table.

| Object | Shape | Prediction | Observation |
|---|---|---|---|
| | | Float ☐<br>Sink ☐ | Float ☐<br>Sink ☐ |
| | | Float ☐<br>Sink ☐ | Float ☐<br>Sink ☐ |
| | | Float ☐<br>Sink ☐ | Float ☐<br>Sink ☐ |
| | | Float ☐<br>Sink ☐ | Float ☐<br>Sink ☐ |
| | | Float ☐<br>Sink ☐ | Float ☐<br>Sink ☐ |

Student's Book p **64**
**4.5** Floating and sinking

# When did it sink?

Draw the boat when it was floating and when it had sunk.

Label your pictures to describe your observations.

| This is the boat when it was **floating**. |
|---|
| This is the boat when it had **sunk**. |

Student's Book p **66**
**4.6** Listen carefully

# What sound does it make?

Draw lines to show the source of each of the sounds.

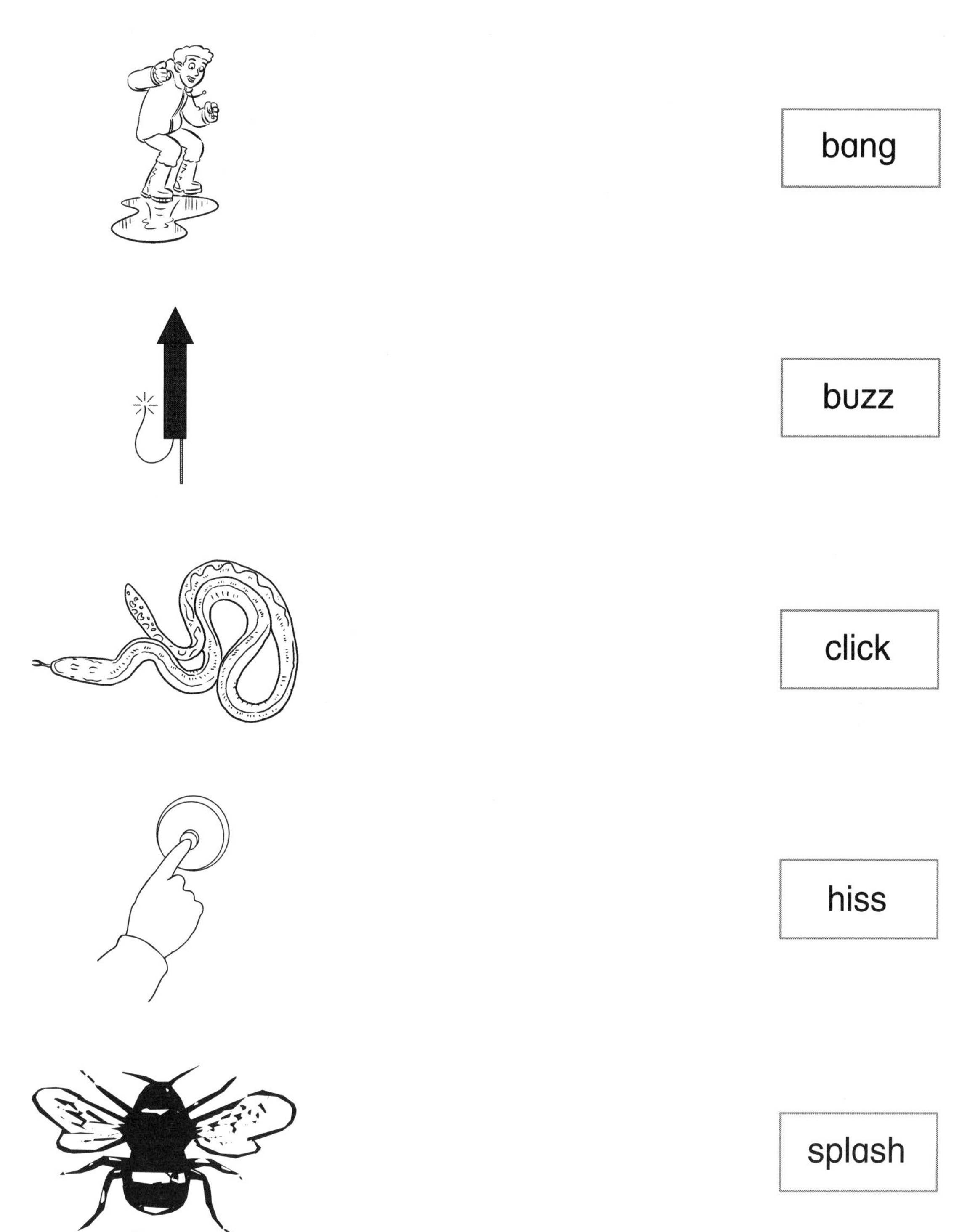

Student's Book p 66
4.6 Listen carefully

# What makes this sound?

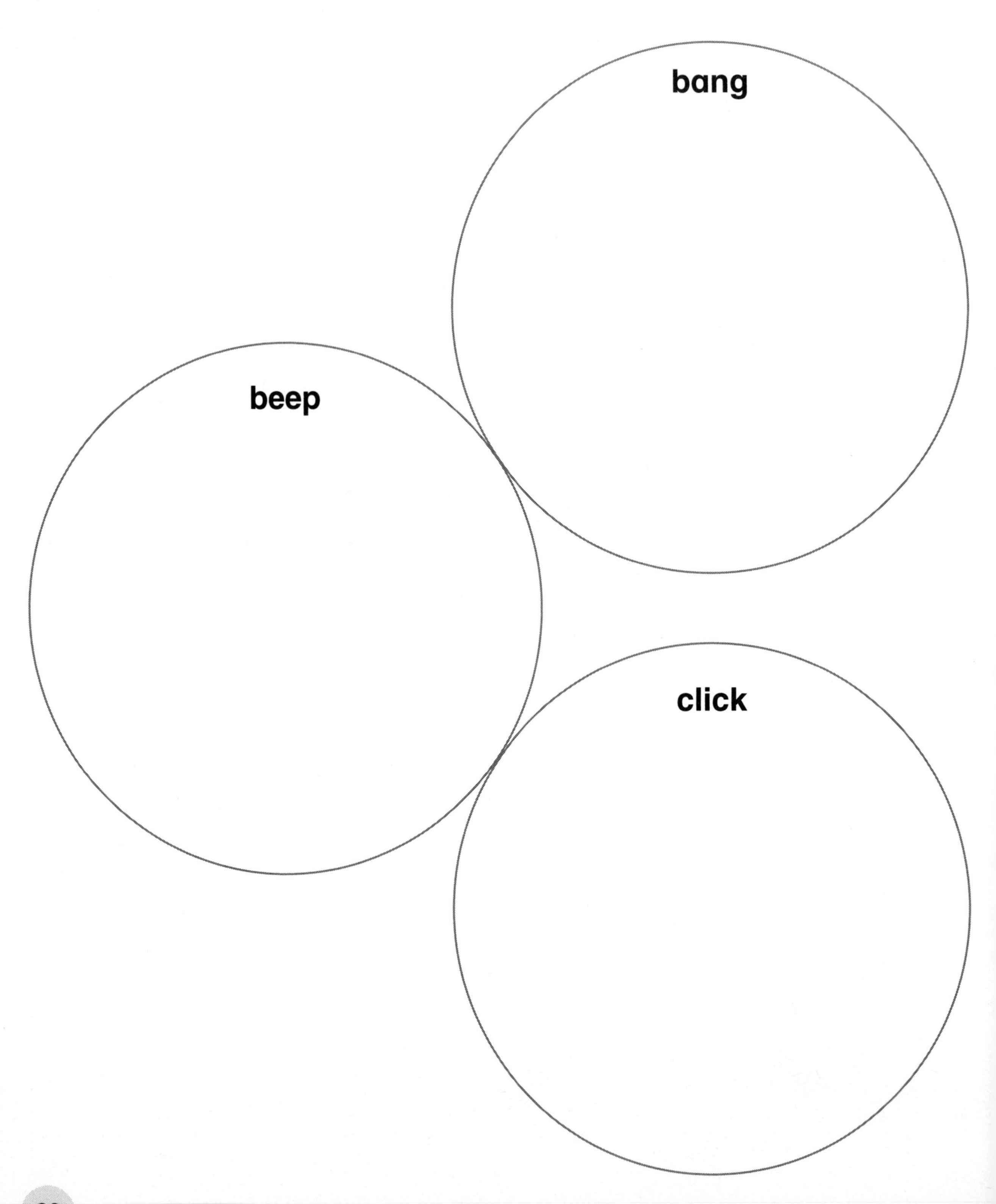

Student's Book p 68
4.7 What made that sound?

# Sounds all around us

Look at the picture. Circle all the sources of sounds.

Student's Book p 70
4.8 Loud and quiet sounds

# Loud or quiet?

**1** Look at the sources of sounds. Draw a line from each source to a number to show how loud the sounds are.

1 = quietest sound 5 = loudest sound

**2** If you moved further away from each source would the sound be louder or quieter?

___

Student's Book p **70**
**4.8** Loud and quiet sounds

# My shaker (1)

Draw and label a picture of your shaker here.

Inside the shaker is ______________________________
______________________________________________.

It sounds ________________________________________
______________________________________________.

Student's Book p 70
4.8 Loud and quiet sounds

# My shaker (2)

Write your plan and predictions here.

I will change the sound that my shaker makes by ______________

______________________________________________

______________________________________________

______________________________________________.

I will try these different things inside my shaker:

______________________________________________

______________________________________________

______________________________________________

I predict the ______________ will make the quietest sound.

I predict the ______________ will make the loudest sound.

I will test how loud or quiet my shaker is:

I predict a person standing close to my shaker will hear a ______________ sound.

I predict a person standing further away from my shaker will hear

a ______________ sound.

Student's Book p 70
4.8 Loud and quiet sounds

# My shaker (3)

Describe your results here.

When I put ______________ inside my shaker it sounded __________

________________________________________________________________.

When I put ______________ inside my shaker it sounded __________

________________________________________________________________.

When I put ______________ inside my shaker it sounded __________

________________________________________________________________.

Compare your results with your predictions here.

The ________________ made the quietest sound.

The ________________ made the loudest sound.

The person standing close to my shaker heard a ________________ sound.

The person standing further away from my shaker heard a ________________ sound.

Look back at your predictions. Were you correct?

Yes ☐ No ☐ Not sure ☐

Student's Book p 72
4.9 Sound and distance

# Can I hear it? (1)

You are going to investigate what happens to a sound as you move further away from the source.

Draw a picture of the source of the sound here.

Describe the sound the source makes.

______________________________________________

______________________________________________

Write your prediction here.

*When I move further away from the source* ______________________

______________________________________________

______________________________________________.

*I think this because* ______________________________

______________________________________________

______________________________________________.

Student's Book p **72**

**4.9** Sound and distance

# Can I hear it? (2)

Draw or write your plan here.

Student's Book p 72
4.9 Sound and distance

# Can I hear it? (3)

Carry out your plan and record your results here.

| Distance from the source | Could I hear the sound? Circle your answer. |
|---|---|
| | Yes No |
| | Yes No |
| | Yes No |
| | Yes No |
| | Yes No |

Draw the positions of you and the source when you could not hear the sound any more.

Student's Book p **72**

**4.9** Sound and distance

# Can I hear it? (4)

Compare your results with your prediction here.

*When I moved away from the source* ____________________

____________________

____________________

____________________

____________________

____________________.

Look back at your prediction. Were you correct?

Yes ☐ No ☐ Not sure ☐

Explain your results here.

____________________

____________________

____________________

____________________

____________________

____________________

Student's Book p 72
4.9 Sound and distance

# Crossword puzzle

How well do you remember the key words from this topic?

Do this crossword puzzle to find out.

### Across

**2** The opposite of a quiet sound is a ______________ sound.

**5** You can blow a whistle to make ______________.

**6** Run, jump and swim are all types of ______________.

**8** As you move away from a sound it gets ______________.

### Down

**1** The place that a sound comes from is called the ______________.

**3** ______________ affects how loud or quiet a sound is.

**4** A push is a type of ______________.

**7** A ______________ is a force that makes an object move towards you.

Student's Book p 74
Looking back Topic 4

# Ball games

Make a list of different ball games.

Write down how you make the ball move in each game.

| Name of ball game | How you make the ball move |
|---|---|
| *football* | *kicked or headed* |
| | |
| | |
| | |
| | |
| | |
| | |

Student's Book p **74**
Looking back **Topic 4**

# Sounds at home

What sounds can you hear at home?

_______________________________________________

_______________________________________________

_______________________________________________

Find the source of each sound and draw a picture of each source of sound.

Student's Book p 76
5.1 What things need electricity?

# Electricity search

Many different objects use electricity.

Did you find these things?

Add some other things that you found.

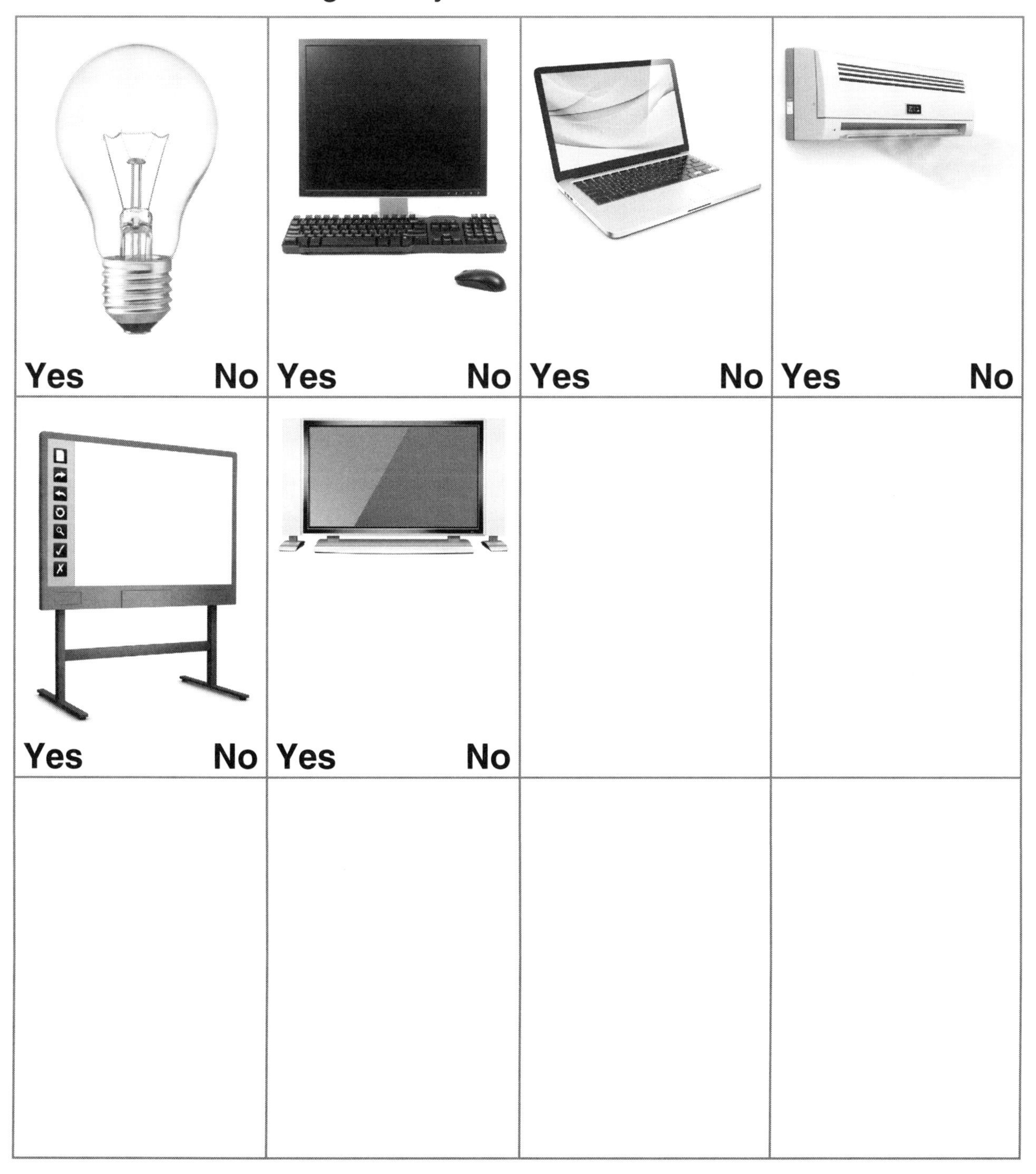

Student's Book p 76
5.1 What things need electricity?

# Things that use electricity

Find four things that use electricity.

Draw and label them.

Describe what each thing does.

Student's Book p 78

5.2 Exploring magnets

# Is it magnetic?

Test some objects to find out if they are magnetic.

Record your findings in the table.

| Object | Is the object magnetic? |
|---|---|
| | |
| | |
| | |
| | |
| | |

How do you know which objects are magnetic?

---

Can you tell if something is magnetic by looking at it?

---

Student's Book p 78
5.2 Exploring magnets

# Magnetic or not?

Find four metal objects and four non-metal objects.

Made a prediction and then test each object to find out if it is magnetic.

Record your findings in the table.

| Object | Material | Prediction | Result |
|---|---|---|---|
| | | Magnetic ☐<br>Not magnetic ☐ | |
| | | Magnetic ☐<br>Not magnetic ☐ | |
| | | Magnetic ☐<br>Not magnetic ☐ | |
| | | Magnetic ☐<br>Not magnetic ☐ | |
| | | Magnetic ☐<br>Not magnetic ☐ | |
| | | Magnetic ☐<br>Not magnetic ☐ | |
| | | Magnetic ☐<br>Not magnetic ☐ | |
| | | Magnetic ☐<br>Not magnetic ☐ | |

Student's Book p 82
Looking back Topic 5

# Electricity at home

Draw pictures of things that use electricity at home.

Add labels to describe what each thing does.

Student's Book p 84
6.1 Clean water investigation

# Filtering water

Watch your teacher make some clean water.

Label the diagram with the names of the equipment.

cotton fabric　cotton wool　elastic band
gravel　plastic bottle　sand

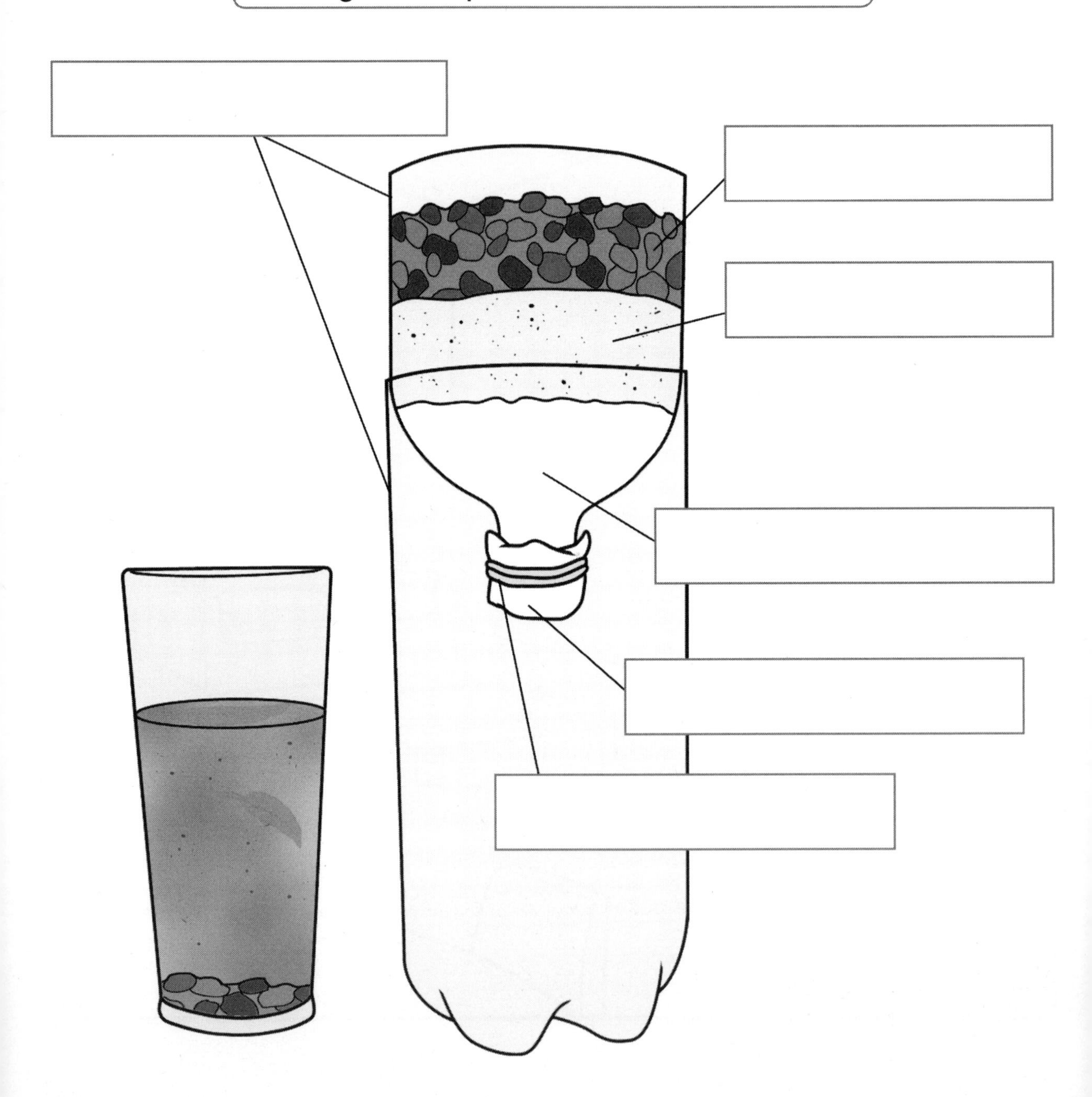

Student's Book p 84
6.1 Clean water investigation

# Testing water filters

*I think the best filter will be* ______________________________.

*I think the worst filter will be* ______________________________.

*I think this because* ______________________________

______________________________________________

______________________________________________.

Draw what the water looked like after it was filtered.

| Cotton ball filter | Paper napkin filter | Coffee filter | Plastic bag filter |
|---|---|---|---|
| | | | |

*I found* ______________________________

______________________________________________

______________________________________________.

Look back at your predictions. Were you correct?

Yes ☐ No ☐ Not sure ☐

Student's Book p **86**
**6.2** Our planet Earth

# Planet Earth

Draw and label the planet Earth.

Complete the sentences using words from the box.

| oceans | Earth | land | water | rivers |
|---|---|---|---|---|

**1** The planet that we live on is called _______________.

**2** Most of the Earth's surface is _______________.

**3** The parts of the Earth's surface that are not water are _______________.

**4** _______________ and _______________ are places where we can find water on Earth.

Student's Book p 88



# Water at home

Think about where and how you use water at home.

Write or draw pictures.

| Where you use water | How you use water |
|---|---|
| bathroom | having a shower |
| | |
| | |
| | |

How can you use less water?

______________________________________________

______________________________________________

Student's Book p **90**

**6.4** What is land made of?

# Rocks, stones and soil

**1** Draw a picture of your favourite rock or stone.

Why did you choose your rock?

*I chose my rock because* ______________________________

______________________________________________.

Words to describe my rock:

____________ ____________ ____________

**2** Land is made from rocks and soil. Investigate a small patch of land at school or home. Draw a picture to show what you can see.

Student's Book p 92
6.5 The Sun

# Light sources

Find some sources of light.

Write where you found each source.

| Light source | Where the light source was |
|---|---|
| | |
| | |
| | |
| | |
| | |
| | |

Draw and label one of the light sources that you found.

Choose words from the box to complete the sentences.

heat space star

1 The Sun is a source of light and ______________.

2 The Sun is a ______________.

3 There are many stars in ______________.

Student's Book p **92**
**6.5** The Sun

# How bright?

Draw or stick your pictures here in order from the most bright to the least bright.

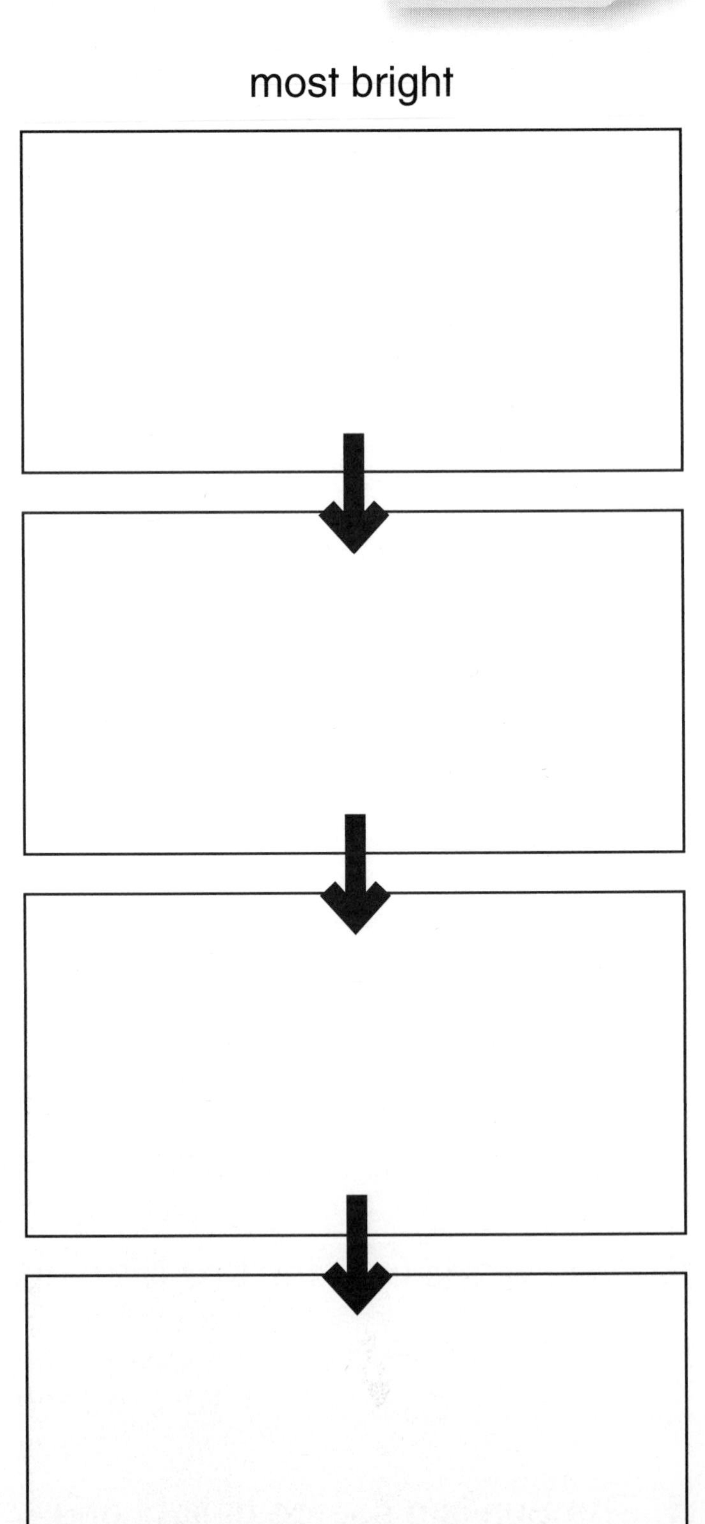